A TRUE BOOK™

FOUR DAYS ON THE TITANIC

Laura McClure Anastasia

Children's Press®
An imprint of Scholastic Inc.

Content Consultant

Tim Maltin, *Titanic* author and historian

Library of Congress Cataloging-in-Publication Data
Names: McClure Anastasia, Laura, author.
Title: Four days on the Titanic/Laura McClure Anastasia.
Other titles: True book.
Description: First edition. | New York: Children's Press, an imprint of Scholastic Inc., 2022. | Series: A true book | Includes bibliographical references and index. | Audience: Ages 8–10. | Audience: Grades 4–6. | Summary: "Next set in A TRUE BOOK series. Young readers rediscover the story of the largest and most luxurious ship ever built, The Titanic. Featuring historical imagery, first-hand accounts, and lively text"—Provided by publisher.
Identifiers: LCCN 2022002394 (print) | LCCN 2022002395 (ebook) | ISBN 9781338840537 (library binding) | ISBN 9781338840544 (paperback) | ISBN 9781338840551 (ebk)
Subjects: LCSH: Titanic (Steamship)—Juvenile literature. | Shipwrecks—North Atlantic Ocean—Juvenile literature. | Ocean travel—Juvenile literature. | Ocean liner passengers—North Atlantic Ocean—Juvenile literature. | Shipwreck victims—Juvenile literature. | Shipwreck survival—Juvenile literature.
Classification: LCC G530.T6 M374 2022 (print) | LCC G530.T6 (ebook) | DDC 910.9163/4—dc23/eng/20220221
LC record available at https://lccn.loc.gov/2022002394
LC ebook record available at https://lccn.loc.gov/2022002395

10 9 8 7 6 5 4 3 2 1 23 24 25 26 27

Printed in China 62
First edition, 2023

Design by Kathleen Petelinsek
Series produced by Spooky Cheetah Press

Front cover: The *Titanic* sinks; (top) a passenger in the gym; (bottom) passengers strolling on the deck

Back cover: The *Titanic* shipwreck was front-page news.

Find the Truth!

Everything you are about to read is true ***except*** for one of the sentences on this page.

Which one is **TRUE**?

T or F The *Titanic* immediately broke in half after hitting the iceberg.

T or F Passengers could pay to send messages from the ship.

Find the answers in this book.

What's in This Book?

A young passenger plays on the top deck.

The gym on the *Titanic* had state-of-the-art equipment, like this rowing machine.

Titanic survivors wait to be picked up by a rescue ship.

INTRODUCTION

The Doomed Voyage

Most people have heard of the ship called the *Titanic* and the tragic tale of its sinking on the night of **April 14–15, 1912**, after hitting an iceberg. The story is famous around the world. But history is filled with shipwrecks, including some that had a much greater loss of life. So why are people still **fascinated by the *Titanic***, more than **100 years after** it sank?

The *Titanic* was world-famous long before it set sail. After all, the *Titanic* and its twin ship, the *Olympic*, were the **largest ships ever built** at that time. And the **most luxurious**.

Many people **believed the *Titanic* could not sink**. But it did—on its very first voyage. Ever since that fateful night, the *Titanic*'s story has continued to grow in people's imagination. This is the story of **what life was like during the four days** the ship was at sea—and what happened to the passengers when it crashed.

In this illustration, the *Titanic* steams across the Atlantic Ocean.

CHAPTER 1

The Voyage Begins

On April 10, 1912, the *Titanic* was ready to set sail from Southampton, England, to New York City. The ship sounded its whistles, rumbling a deep farewell as it set off. The *Titanic*'s first stop was in France, where it welcomed additional passengers. Then it sailed to Ireland, where more people boarded the next morning. On Thursday, April 11, the mighty ship set a course for New York City. Experts estimate that more than 2,200 people were aboard for the ship's first voyage.

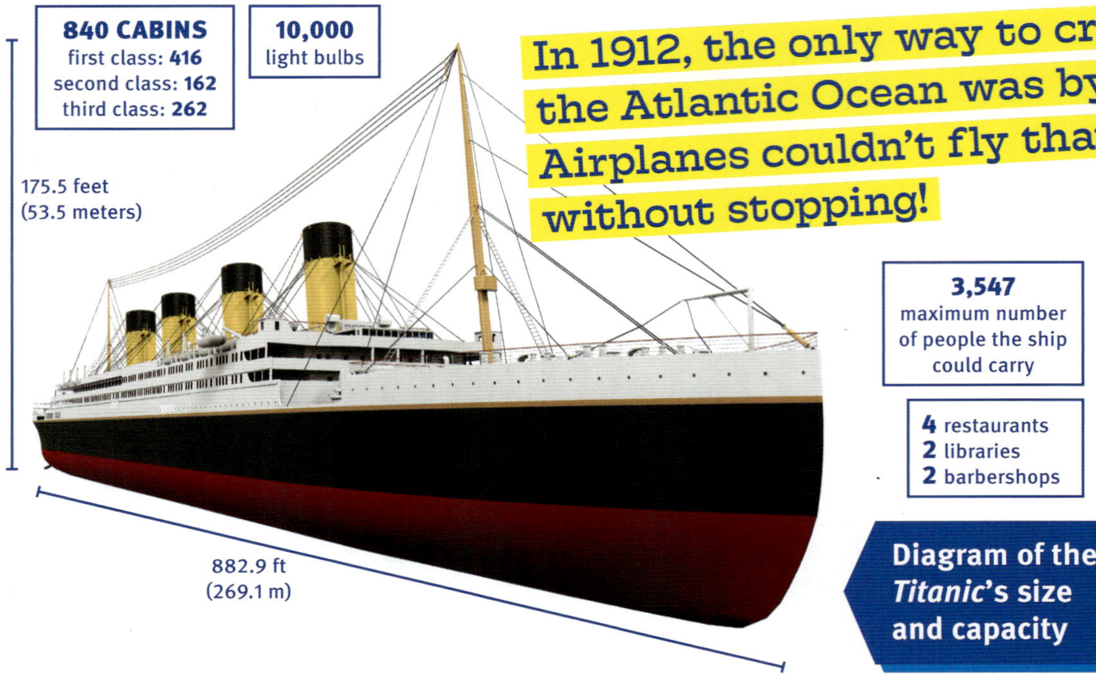

Diagram of the *Titanic*'s size and capacity

In 1912, the only way to cross the Atlantic Ocean was by ship. Airplanes couldn't fly that far without stopping!

The World's Biggest Ship

Much of the excitement surrounding the trip was due to the *Titanic*'s size. The **ocean liner** was about two-and-a-half football fields long and weighed 46,328 tons. That was about 15,000 tons heavier than its competitors. The ship also had an inventive design. Its **hull** was divided into 16 watertight compartments. Engineers said the ship could stay afloat even if four of those compartments flooded. Many people thought the *Titanic* was unsinkable.

Separated by Class

As the *Titanic* set off across the Atlantic Ocean that Thursday afternoon, passengers gathered for lunch in the dining rooms. Then they explored the ship. Passengers were divided into three classes based on the cost of their ticket. Travelers in each class had to stay in their own part of the ship. Every **stateroom** had heat, electric lights, and working sinks. But only first-class passengers could access most of the ship's special features, including a gym, a heated swimming pool, and a spa known as a Turkish bath.

The *Titanic* was one of the first ships with a heated swimming pool.

Parlor suites, the most expensive cabins, had sitting rooms like this one.

The Finest Cabins

The *Titanic*'s richest passengers bought first-class tickets. Crew members reportedly greeted all 324 of them personally as they boarded, then showed them to their staterooms. First-class cabins were on the highest decks and near the center of the ship. Those areas were farthest from the water and offered the smoothest sailing. The staterooms were richly decorated with hand-carved furniture. The best of those were parlor suites. Each one had two bedrooms, a sitting room, and a private bathroom—and two of them had private decks.

Second-Class Style

The 284 passengers who traveled in second class included teachers, church leaders, and the servants of first-class passengers. Their staterooms were mostly on the middle decks of the ship. Most of the staterooms had bunk beds, a sofa, and wooden furniture. The second-class cabins weren't as luxurious as the *Titanic*'s more expensive ones. They were more luxurious than first-class staterooms on other ships, though.

A second-class cabin

There were only two bathtubs to be shared by everyone in third class!

Third-class cabins on the *Titanic* were small and simple.

Third-Class Travel

The greatest number of passengers traveled in third class. There were 709 of these passengers, and each had to be examined by a medical officer before boarding. The doctor wanted to make sure they were not carrying any diseases to America. Many third-class passengers were **emigrants** from Europe and Asia who hoped to find a better life in the United States. Their cabins were on the lowest decks, closest to the ship's engines. The simple but comfortable third-class rooms could sleep up to six people.

No Women Allowed

People on the *Titanic* were separated by more than wealth—or lack of it. First-, second-, and third-class areas each had a men-only smoking room where male passengers could smoke and play cards. At the time, it was considered improper for women to smoke in public, so they were not allowed in these special rooms. However, there was a reading and writing room in first class that was mainly used by female passengers. Wealthy women headed there to read, write letters, and talk.

First-class smoking room

First-class reading and writing room

In first class, dinners could last four hours!

The *Titanic*'s Café Parisien (pictured) was designed to look like a café in Paris!

CHAPTER

2

A Vacation at Sea

By Friday, April 12, two days after leaving England, the *Titanic*'s passengers were in full vacation mode. People strolled the ship's decks, socialized with new friends, and played music and games. And when the passengers weren't relaxing, they were eating. Most meals were included with the price of a ticket. Breakfast was served from 8 a.m. to 10 a.m., lunch was offered at 1 p.m., and the evening meal started around 7 p.m. Each class had its own menus and dining **saloon**.

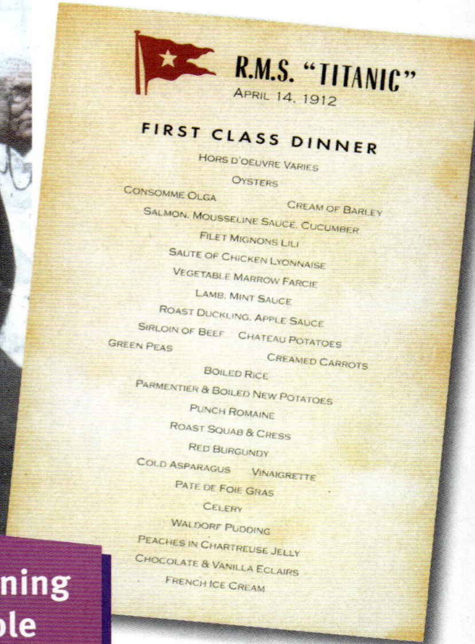

R.M.S. "TITANIC"
April 14, 1912

FIRST CLASS DINNER

Hors D'Oeuvre Varies
Oysters
Consomme Olga — Cream of Barley
Salmon, Mousseline Sauce, Cucumber
Filet Mignons Lili
Saute of Chicken Lyonnaise
Vegetable Marrow Farcie
Lamb, Mint Sauce
Roast Duckling, Apple Sauce
Sirloin of Beef — Chateau Potatoes
Green Peas — Creamed Carrots
Boiled Rice
Parmentier & Boiled New Potatoes
Punch Romaine
Roast Squab & Cress
Red Burgundy
Cold Asparagus — Vinaigrette
Pate de Foie Gras
Celery
Waldorf Pudding
Peaches in Chartreuse Jelly
Chocolate & Vanilla Eclairs
French Ice Cream

The first-class dining room and a sample dinner menu

Fancy Feasts

Meals were an elegant affair in first class. The *Titanic*'s beautifully decorated dining saloon was the biggest of any ship. Wealthy passengers dressed in their fanciest clothes and jewels. The orchestra played while passengers dined on 10-course meals that included steak and roast duck. First-class passengers could also pay to eat at a French-style restaurant. And they could get small refreshments throughout the day at two cafés overlooking the water.

Food for Days

The trip from Europe to New York should have taken five or six days. The *Titanic*'s kitchen crew needed a lot of food to serve thousands of meals every day. Here are some of the ingredients and their quantities that the big ship was probably carrying:

- 80,000 pounds (36,000 kilograms) of potatoes
- 75,000 pounds (34,000 kg) of fresh meat
- 40,000 eggs
- 16,000 lemons
- 10,000 pounds (4,500 kg) of sugar
- 6,000 pounds (2,700 kg) of butter
- 1,500 gallons (5,600 liters) of milk
- 1,200 to 1,750 quarts (1,100 to 1,600 l) of ice cream
- 200 barrels of flour

Everything needed for a ship's journey had to be placed on board before the ship set sail.

Simpler Meals

Second-class passengers also dined in a large dining saloon, where a pianist entertained them while they ate. The menu included roast turkey and lamb, but offered fewer courses than in first class. Third-class passengers ate more basic foods, like meat, potatoes, and bread. But much of the food was better than many would have enjoyed on land. The third-class dining hall was not large enough to fit all the passengers at once, so they ate in shifts.

White Star Line

Triple Screw Steamer "Titanic."

2ND. Class

April 14, 1912.

DINNER.

Consommé Tapioca

Baked Haddock, Sharp Sauce

Curried Chicken & Rice

Spring Lamb, Mint Sauce

Roast Turkey, Cranberry Sauce

Green Peas | Purée Turnips

Boiled Rice

Boiled & Roast Po

Plum Pudding

Wine Jelly | Cocoan

American Ice Cr

ts Assort

esh Frui

Bi

Coffee

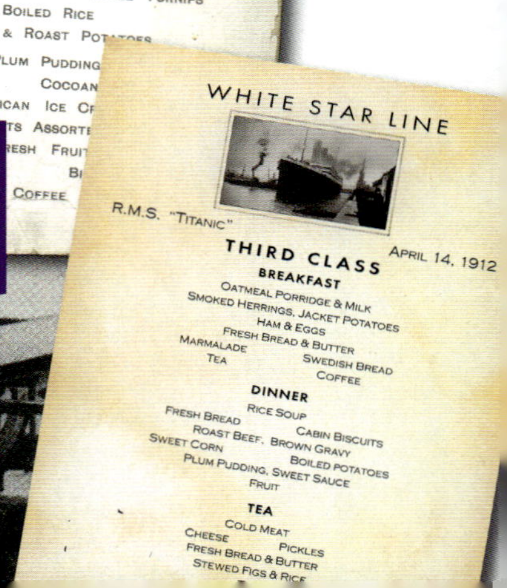

WHITE STAR LINE

R.M.S. "Titanic" | April 14, 1912

THIRD CLASS

BREAKFAST

Oatmeal Porridge & Milk

Smoked Herrings, Jacket Potatoes

Ham & Eggs

Fresh Bread & Butter

Marmalade | Swedish Bread

Tea | Coffee

DINNER

Rice Soup

Fresh Bread | Cabin Biscuits

Roast Beef, Brown Gravy

Sweet Corn | Boiled Potatoes

Plum Pudding, Sweet Sauce

Fruit

TEA

Cold Meat

Cheese | Pickles

Fresh Bread & Butter

Stewed Figs & Rice

On other ships, third-class passengers had to bring their own food.

The third-class dining room and menus from second and third class

There was no gift shop on the *Titanic*. The barbershops sold some souvenirs and *Titanic*-themed gifts.

Souvenirs hang from the ceiling of this barbershop on the *Titanic*.

Everyday Convenience

By Saturday morning, the third full day of the trip, the *Titanic* was in the middle of the Atlantic Ocean. Still, passengers were able to keep to their regular daily routines. The ship offered church services, as well as two barbershops, a Sea Post Office, and even a hospital with two doctors. For passengers who wanted to keep up with news from home, there was an on-board newspaper called the *Atlantic Daily Bulletin*. The paper ran articles, horse-race results, the ship's daily menus, and more.

The gym was used by men and women at different times of the day.

This man is using a rowing machine in the *Titanic*'s gym.

Daytime Entertainment

Shortly after noon on Saturday, April 13, the crew posted the run. That is the distance the ship had traveled over the previous 24 hours. Passengers eagerly bet on what the next run might be. There were plenty of other ways to stay entertained as well. First-class passengers swam in the pool, exercised in the gymnasium, and relaxed in the Turkish bath. Second-class passengers wrote letters and read books in their section's library. People in all three classes played games and strolled on their decks.

Fun for Kids

Most of the 100-plus children aboard the *Titanic* were in third class. By Saturday morning they had come up with their own ways to pass the time. The handful of children in first class had built-in fun. They could use the ship's playground. And they were allowed to visit the gym at certain times. They could ride on mechanical saddles that rocked back and forth to imitate riding on a horse or a camel. Kids in second and third class raced on the decks and played made-up games. Others explored the ship.

A children's playground (pictured) was on the top deck.

Early Warnings

The temperature dropped steadily on Sunday, April 14. Over the course of the day, the *Titanic*'s radio operators received warnings from other ships about icebergs in the water. The messages, which became more frequent as the day went on, came through a wireless radio system. The wireless operators communicated with other ships—and people on land—using **Morse code**. John Edward Smith was the *Titanic*'s captain. In light of the warnings, he adjusted the ship's course slightly around 6 p.m.

Passengers could pay to have messages sent out through the wireless radio system.

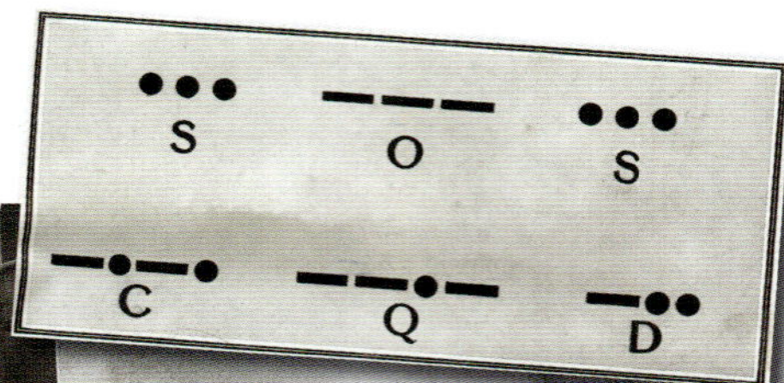

Wireless operators sent messages using Morse code, shown here. Letters are made from a series of dashes and dots.

The band took requests, so its musicians had to know hundreds of songs by heart.

These are some of the musicians who were on the *Titanic*.

Evening Entertainment

After dinner, the first-class passengers gathered in their reception room to listen to the band play, as they had done the other nights. Many second-class passengers held a sing-along in their dining saloon. Meanwhile, the third-class passengers threw a party in their common area, playing instruments and dancing. By 11 p.m., many passengers had returned to their cabins. But the ship's **lookouts** were on high alert, watching for icebergs.

8 THE DAILY GRAPHIC SPECIAL TITANIC IN MEMORIAM NUMBER, APRIL 20, 1...

THE OCEAN GRAVE OF THE TITA...

WORLD-WIDE HELP.

FUNDS FOR THE RELIEF OF SUFFERERS.

MANSION HOUSE FUND.

KING AND QUEEN AMONG THE SUBSCRIBERS.

Widespread distress will inevitably follow upon a disaster of such magnitude as the sinking of the Titanic, and all the money which can be collected will be required to alleviate the suffering.

The Lord Mayor (Sir T. B. Crosby) on Wednesday opened a Mansion House Fund, and below we publish a letter from Sir Thomas in which he appeals to the generosity of the nation.

Among the first to contribute to the fund were their Majesties, Queen Alexandra, and the Princess Royal. These donations were as follows:—

The King	500 gns.
The Queen	250 gns.
Queen Alexandra	£200
The Princess Royal	£100

By Thursday night the Fund amounted to over £25,000, and each succeeding post brought further donations to swell the total. The larger of the first subscriptions included the following:—

Shipping Federation...£2,100; Messrs. Morgan, Grenfell and Co. ... 2,000; Corporation of London 1,050; Messrs. Speyer Bros. ... 1,050; Lord Mount Stephen ... 1,000; Sir William Nelson ... 1,000; Mr. Edward C. Grenfell ... 1,000; Messrs. N. M. Rothschild and Sons ... 525; Canadian Agency, Ltd. 525; Messrs. J. Henry Schroder and Co. ... 525; Messrs. Lazard Bros. and Co. ... 525; Baring Bros. and Co., Ltd. ... 525; Messrs. Baker, Mason, and Co. ... 500; Messrs. Brown, Shipley, and Co. ... 500; Messrs. Emile Erlanger and Co. ... 500; Messrs. Glyn, Mills, Currie and Co. ... 500; Messrs. George Kitchin and Co. ... 500; The American Ambassador (Mr. Whitelaw Reid) ... 500; Messrs. C. Hambro and Son ... 500; Royal Mail Steam ... 500

...PEAL.

...following

...APHIC.

...timate that ...nsion House ...relief of the ...tives of those ...ers—who have ... calamity, and ...nce of the benevolent public ... to invite the ... to relieve, in some degree, the distress which has been occasioned in many hundreds of families by a disaster fortunately unparalleled in the history of ocean navigation.

In taking this step I feel sure that I am promptly responding to the wishes of those who urge that the keen sympathy, universally and unstintedly entertained for those who have thus suddenly been plunged into misery and distress, should assume some practical shape for the future advantage of the bereaved families.

Of the real extent of the calamity it is too soon to expect reliable details. Some time must necessarily elapse before information can be obtained as to the number of those lost and their wives and families and their circumstances, but sufficient is known to make it evident that a very large sum ... adequately provide for those in ... the Work...

PERIL OF THE BERG.

FLOATING ICE-FIELDS THAT HARASS THE NAVIGATOR.

WHAT CAN BE DONE?

It was the misfortune of the Titanic to begin her career at a moment when the North Atlantic was exceptionally dotted with icefields and icebergs. There is a season for Atlantic ice, as for everything else. It opens with the month of April. It closes with the month of August. Anywhere within this period you may have much ice or little.

One ice season is never just like another. The present has opened badly from the navigator's point of view. For reasons which are difficult to dogmatise about, but which probably are the result of climatic eccentricity, the ice has come down in the very early days of April, driven irresistibly by currents to the warmer waters into which it is ultimately to disappear.

It is curious that, just as the Atlantic now appears at the beginning of the season to be at its worst so far as ice is concerned, it was at the end of last season that the ice danger was most marked.

On August 6th of last year the Anchor Line steamer Columbia arrived at New York with her bow plates crushed in for a distance of fifteen feet.

NOTABLE TRIBUTES.

MESSAGES FROM THE KING AND THE KAISER.

PREMIER'S ELOQUENCE.

It is fitting that a disaster so tremendous as the sinking of the Titanic should have called forth notable tributes of grief and sympathy.

Chief among these are the messages of the King. To the White Star Line he telegraphed:—

The Queen and I are horrified at the appalling disaster which has happened to the Titanic and at the terrible loss of life.

We deeply sympathise with the bereaved relations, and feel for them in their great sorrow with all our hearts. GEORGE R. and I.

His Majesty's feeling message to President Taft was worded as follows:—

The Queen and I are anxious to assure you and the American nation of the great sorrow which we experience at the terrible loss of life that has occurred among the American citizens and my own subjects by the foundering of the Titanic.

Our two countries are so intimately allied by ties of friendship and brotherhood that any mis... fortune which affects the one must necessarily affect the other, and on the present heartrending occasion they are both equally sufferers.

GEORGE R. and I.

(Signed) ... to the White Star ...

DOOMED...

LAST MESSA... BE...

'MANY T...

Tragic inter... which several ... the Titanic, ... out by the lin...

The French ... Havre on Mo... of the Titan... the ice, an... captain.

At midnig... latitude 44... the Tourain... duced in ... o'clock in th... ice was ve... of the sam... edge of an... until 6.45. ...

The Tour... munication ... in the eve... the comm... position o... who repli...

"GOO...

The C... Queenstow... ported ... from the ... Star line ... was in ... diate ... from th... Caronia ... time to ... Anoth... on ar... Canada ... spoke ... sage: "Many ... The ... entere... fully ... stopp... miles ... seen. ... thirty...

SENDER OF THE "S. O. S." SIGNAL FOR HELP.

Mr. G. G. Phillips, the Marconi wireless operator on board the Titanic, whose signal, "S.O.S.," was received by liners hundreds of miles away. (Photographed by Jennie Stedman.)

It is said that her port anchor, which was lost, was ... into which she crashed. She ... the Donald...

This newspaper ran a special "in memoriam" feature on the *Titanic*. Radio operator Jack Phillips is pictured.

CHAPTER 3

The Crash

At around 10:55 p.m., Jack Phillips, one of the *Titanic*'s radio operators, received what would be the *Titanic*'s final warning. The *Californian*, a ship that was about 20 miles (32 kilometers) away from the *Titanic*, messaged that it was surrounded by ice. Phillips was struggling to catch up on passenger messages. He dismissed the alert and told the radio operator on the *Californian* to stop sending him messages so he could get his work done. The radio operator on the *Californian* went to bed.

Iceberg Spotted

Meanwhile, the two lookouts in the *Titanic*'s **crow's nest** continued scanning the dark sea for icebergs. At about 11:40 p.m., they spotted a mass of ice suddenly appearing against a strange haze on the horizon. One of the lookouts rang the bell three times to warn the officers steering the ship of an object dead ahead. They also telephoned the **bridge** and informed them, "Iceberg dead ahead!"

The first officer told the crew to steer hard to the left, the first part of a maneuver designed to avoid the iceberg.

Lookouts were posted in the crow's nest, 50 feet (15 meters) above the *Titanic*'s deck.

Calm seas helped hide the iceberg because waves weren't breaking on it.

Thomas Andrews (top) knew the *Titanic* was doomed soon after it hit the iceberg.

Deadly Crash

Then the order was given to reverse the engines to stop the ship. But it was too late. The ocean liner's right side and hull bottom scraped along the iceberg underwater. Water started gushing into the lowest parts of the ship. Crew members assured passengers that everything was under control. As a precaution, Captain Smith ordered the lifeboats to be readied. The *Titanic*'s designer, Thomas Andrews, went belowdecks to survey the damage. He quickly realized the *Titanic* was doomed.

Evacuation

This photo shows how the *Titanic*'s lifeboats were lowered over the side of the ship.

Captain Smith told the crew to start loading the lifeboats—women and children first. Then he directed the radio operators to tap out distress signals asking for help from other ships. Many passengers didn't believe the *Titanic* would sink. They refused to board the lifeboats. But by 1:50 a.m., it became clear that the *Titanic* was going down. Most of the women and children in first and second class made it onto the lifeboats. But many third-class passengers did not even reach the loading area until after the lifeboats had been lowered away.

Missed Signals

As the *Titanic* started to fill with water, some passengers spotted the lights of another ship nearby. They hoped the ship might come to their rescue. But it did not respond to the wireless operators' calls for help. It was the *Californian*, whose radio operator had gone to bed. The *Titanic*'s Officer Joseph Boxhall had shot a **flare** into the sky when the ship started to sink. He fired several more over the next hour or so. But the crew on the *Californian* did not recognize the flares as distress signals. (Ships fired flares for a variety of reasons.) The next morning, when the *Californian*'s radio operator signed on, the *Titanic*'s messages came through. But by the time the *Californian* reached the area where the *Titanic* had been, it was too late.

The BIG Truth

Room for More

The *Titanic* had 20 lifeboats that could hold 1,178 people in all—far fewer people than were aboard the ship. Even with limited lifeboat spots, most of the ship's rescue craft were launched before they were full. The *Titanic*'s officers were worried about overloading the lifeboats, and they wanted to get them away from the ship quickly, before the *Titanic* sank. But that was just part of the reason why only about 700 people made it into the lifeboats. Here are some others.

Disbelief

For at least one hour after the collision, crew members assured passengers that the damage was minor. Many people decided they would be safer on the giant ship than on a small lifeboat in the dark, near-freezing water. It wasn't until most of the lifeboats were gone that many people realized the seriousness of the situation. By then it was too late.

Family

The crew turned many men away from the lifeboats. They thought men were expected to put women's and children's lives ahead of their own. Some women refused to leave their husbands or older sons. One passenger declared she would rather die with her husband than live without him.

Boat Issues

Some crew members worried that full lifeboats would be too heavy to lower over the side of the ship. So they loaded fewer people than the boats were built to hold. Then the crew ran out of time to launch the last two boats. The ocean swept one overboard at around 2:05 a.m., as the bow, or front, of the *Titanic* tilted lower into the ocean. About a dozen people managed to survive in that boat, even though it was partially filled with icy water. The ocean washed the other boat into the sea upside down. About 20 people survived on top of it.

The temperature of the water was 28 degrees Fahrenheit (-2 degrees Celsius), about as cold as an ice cube.

The front part of the *Titanic* broke off and plunged into the ocean.

CHAPTER 4

Lost at Sea

The lifeboats were gone, and about 1,500 passengers and crew members remained trapped on the sinking ship. The band continued to play in an effort to comfort the panicked passengers. Then the weight of the water on the front of the vessel became too much. The *Titanic* broke thunderously in half. At about 2:20 a.m., two hours and 40 minutes after the collision, the mighty ship sank to the seafloor. Some people were flung into the icy ocean. Others jumped in and tried to swim toward the lifeboats—their last hope for survival.

Cries for Help

The water was so cold that the hundreds of people in the ocean were immediately in grave danger. Their cries filled the dark night. But most of the passengers in the lifeboats were too afraid to row closer to help. They feared that if too many people got into the boats, they would **capsize**, endangering everyone on board. Only one lifeboat returned in time to rescue three people. The freezing water made almost everybody else unconscious in about 20 minutes.

The *Titanic*'s Final Hours

10:55 p.m., April 14
The *Titanic* receives its final warning about icebergs in the area.

11:40 p.m., April 14
A lookout spots the iceberg. Less than a minute later, the *Titanic* collides with it.

12:05 a.m., April 15
The captain orders the crew to prepare the lifeboats.

Rescue

People in the lifeboats might have shined lights or lit tiny fires to make themselves visible in the darkness.

For almost two more hours, the small boats floated in the ocean. The air was freezing, and some passengers were wearing little more than pajamas and a life vest. About 4:30 a.m., a ship called the *Carpathia* arrived and began collecting survivors from the scattered lifeboats. The ocean liner had traveled about 58 miles (93 kilometers) through dangerous waters to come to the rescue. For about four hours, the *Carpathia*'s crew helped the survivors climb onto their ship to safety.

12:45 a.m., April 15
The first lifeboat is lowered. It is less than half full.

2:20 a.m., April 15
The *Titanic* plunges beneath the water.

4:30 a.m., April 15
The *Carpathia* arrives to rescue survivors in lifeboats (seen here).

Reaching New York

Aboard the *Carpathia*, the survivors were given warm blankets, dry clothes, and hot food and drinks. Some of the survivors were treated for **frostbite**, shock, and exposure to the icy water. The *Carpathia*'s passengers shared or gave up their cabins to the *Titanic* passengers. It took the *Carpathia* crew four hours to **navigate** the field of ice. Then they continued on for three days before reaching New York City, completing the voyage that the *Titanic* never would.

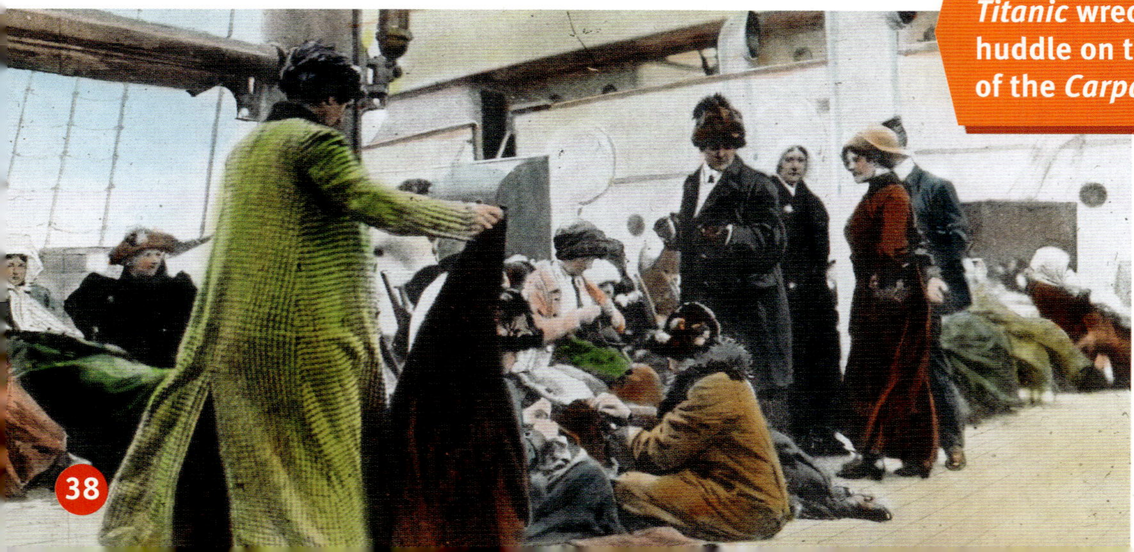

Survivors of the *Titanic* wreck huddle on the deck of the *Carpathia*.

The World Mourns

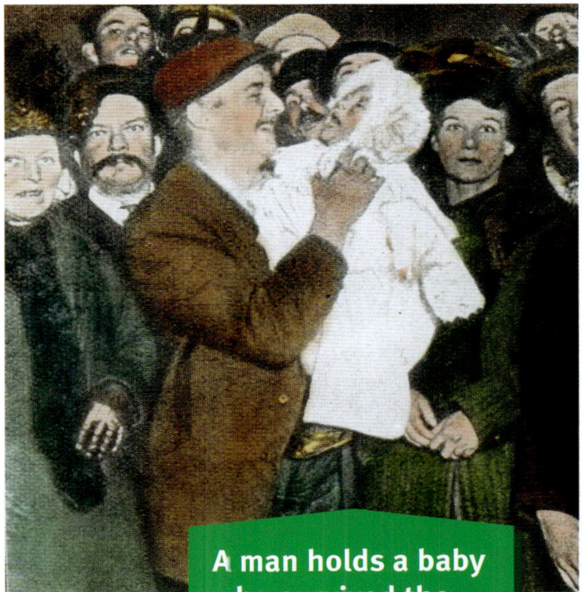

A man holds a baby who survived the shipwreck.

At 9:25 p.m. on April 18, the *Carpathia* slowly approached New York Harbor. Thousands of people were gathered at the docks in New York City awaiting the ship. Many were desperate to find out whether their loved ones who had been on the *Titanic* had survived. But the odds were not in their favor. Only about 713 of the more than 2,200 passengers and crew had made it. The loss of so many lives—aboard the "unsinkable" *Titanic*, no less—shocked the world. To this day, the shipwreck remains one of modern history's most famous tragedies. Museums, movies, and books like this one help keep the *Titanic*'s story alive.

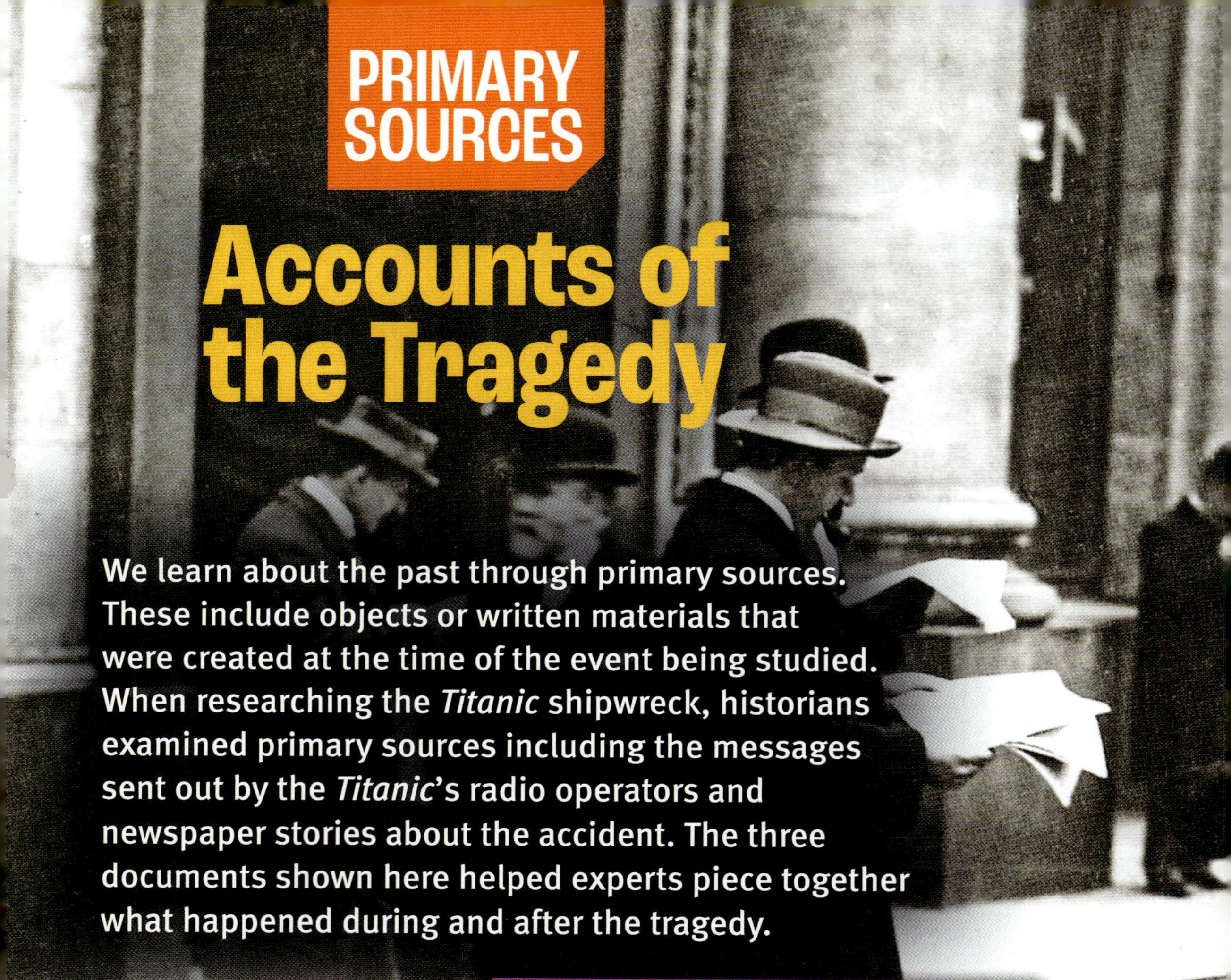

PRIMARY SOURCES

Accounts of the Tragedy

We learn about the past through primary sources. These include objects or written materials that were created at the time of the event being studied. When researching the *Titanic* shipwreck, historians examined primary sources including the messages sent out by the *Titanic*'s radio operators and newspaper stories about the accident. The three documents shown here helped experts piece together what happened during and after the tragedy.

The *Titanic*'s Distress Message

This is the text of the message sent out by the *Titanic*'s radio operator after the ship hit the iceberg. It was written down by the radio operator on another ship. It reads:
SOS SOS CQD CQD
We are sinking fast. Passengers being put into boats.
The individual letters "CQD" and "SOS" were distress codes used to show that a ship needed help.

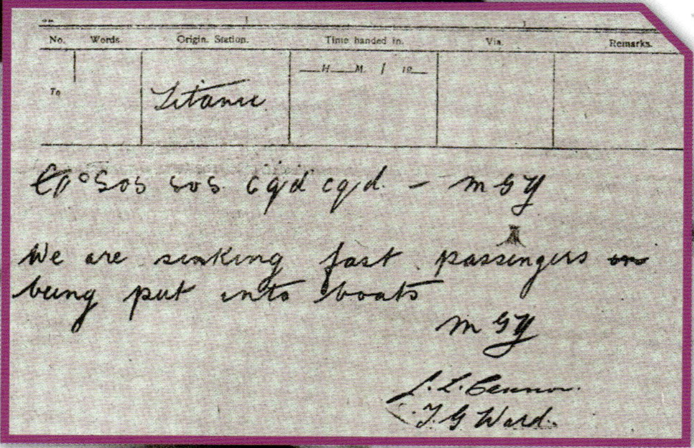

No.	Words.	Origin. Station.	Time handed in.	Via.	Remarks.
To		Titanic	—H.—M. / 19—		

CQD SOS SOS cqd cqd - MGY

We are sinking fast passengers being put into boats

MGY

Front Page of the *Evening Sun*, April 15, 1912

This Baltimore newspaper rushed to print with an inaccurate report. In addition to claiming that everyone had survived the crash, the newspaper reported that the steamship *Virginian* was towing the *Titanic* to Halifax, Nova Scotia, Canada. This was not the only newspaper to report the accident incorrectly.

THE EVENING SUN

VOLUME IV—NO. 154. BALTIMORE, MONDAY, APRIL 15, 1912. 14 PAGES PRICE ONE CENT

Last Edition

ALL TITANIC PASSENGERS ARE SAFE; TRANSFERRED IN LIFEBOATS AT SEA

PARISIAN AND CARPATHIA TAKE HUMAN CARGO

Steamship Virginian Now Towing Great Disabled Liner Into Halifax

ALL DOUBT AS TO STEAMER REACHING PORT SET AT REST

Wireless Messages State That No Lives Have Been Lost On The Damaged Vessel.

WRECK OF TITANIC IS UNPARALLELED

WOMEN AND CHILDREN IN LIFEBOATS

SPECIAL TRAINS FOR WRECK VICTIMS

FORMER MISS POLK IS A PASSENGER

Bulletins

ATLANTIC OCEAN

"All the News That's Fit to Print."

The New York Times.

NEW YORK, TUESDAY, APRIL 16, 1912

TITANIC SINKS FOUR HOURS AFTER HITTING ICEBERG; 866 RESCUED BY CARPATHIA, PROBABLY 1250 PERISH; ISMAY SAFE, MRS. ASTOR MAYBE, NOTED NAMES MISSING

Col. Astor and Bride, Isidor Straus and Wife, and Maj. Butt Aboard.

"RULE OF SEA" FOLLOWED

Biggest Liner Plunges to the Bottom at 2:20 A.M.

RESCUERS THERE TOO LATE

The Lost Titanic Being Towed Out of Belfast Harbor.

PARTIAL LIST OF THE SAVED.

CAPT. E. J. SMITH

Front Page of the *New York Times*, April 16, 1912

This article, published the day after the accident, recounts how the *Carpathia* picked up *Titanic* survivors, and gives realistic numbers of lives lost. With this and other reports, the public learned the grim truth.

DEEPER DIVE

The *Titanic*'s

1

11:41 p.m., April 14

The *Titanic* hits the iceberg.

2

1:50 a.m., April 15

The bow, or front, of the ship is underwater. The pressure causes the first smokestack to break off. More water rushes into the opening.

4

2:18 to 2:20 a.m., April 15

Very quickly, the bow breaks free and starts to sink. Water rushes into the stern, which sinks minutes later.

Final Hours

3

2:18 a.m., April 15

The ship's bow is filled with water and has gone beneath the surface of the ocean. But the stern, or back, is still afloat. The pressure causes the ship to crack in half.

5

Final Resting Place

Experts think it took 5 to 10 minutes for the bow and stern to hit the ocean floor.

Titanic's Bow

The front end of the ship slammed into the ocean floor and settled there.

Titanic's Stern

As the back end of the ship sank, different parts broke free or fell out of the open end of the stern. The weakened stern then hit the ocean floor hard enough to break apart.

True Statistics

Estimated number of passengers and crew aboard the *Titanic*: 2,225

Age of the youngest passenger: 2 months (she survived)

Number of lookouts on the *Titanic*: six, who worked in shifts around the clock

Water temperature when the *Titanic* sank: 28°F (-2 °C), about as cold as an ice cube

Number of lifeboats: 20

Number of people the *Titanic*'s lifeboats could have held: 1,178

Number of passengers who survived: about 705

Minutes it took for the *Titanic* to sink: 160

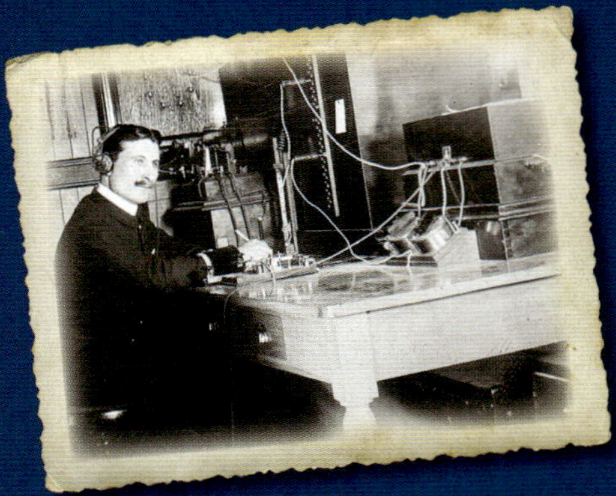

Did you find the truth?

F **The *Titanic* immediately broke in half after hitting the iceberg.**

T **Passengers could pay to send messages from the ship.**

Resources

Other books in this series:

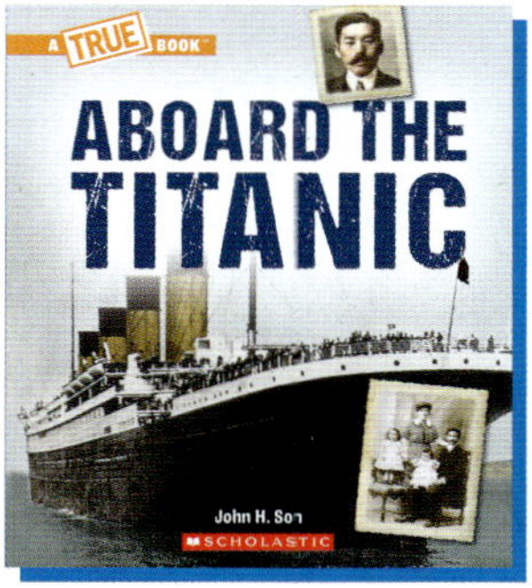

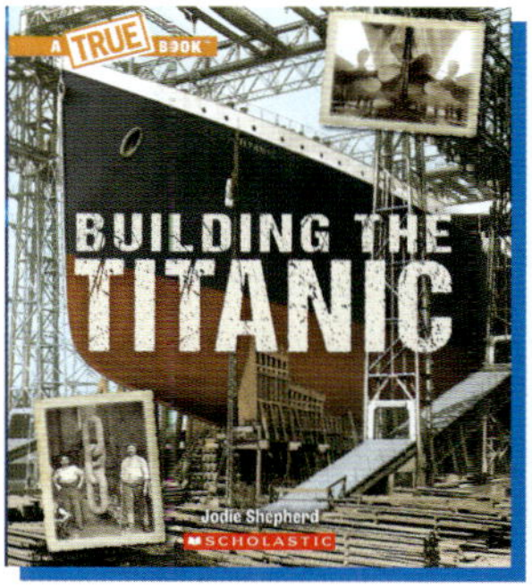

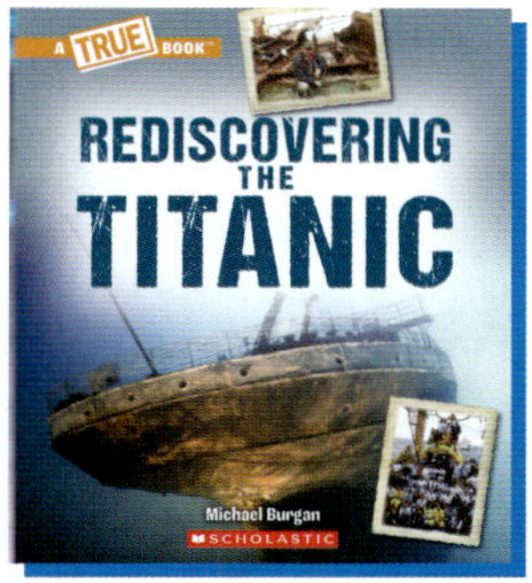

You can also look at:

Benoit, Peter. *The Titanic.* New York: Scholastic, 2013.

Hopkinson, Deborah. *Titanic: Voices from the Disaster.* New York: Scholastic, 2012.

Montero, Mary. *Titanic Q&A: 100+ Fascinating Facts for Kids.* Emeryville, CA: Rockridge Press, 2020.

Oachs, Emily Rose. *The Titanic: Digging Up the Past.* New York: Scholastic, 2020.

Tarshis, Lauren. *I Survived the Sinking of the Titanic, 1912.* New York: Scholastic, 2010.

Glossary

bridge (brij) the ship's command center

capsize (KAP-size) to turn over in the water

crow's nest (KROHZ nest) a small platform used for a lookout, found on a sailing ship's mast

emigrants (EM-i-gruhnts) people who leave one country to settle in another

flare (flair) a flame or bright light used as a signal for help

frostbite (FRAWST-bite) a condition that occurs when extremely cold temperatures damage parts of a person's body, such as fingers, toes, ears, or nose

hull (huhl) the frame or body of a boat or ship

lookouts (LUK-outs) people who keep watch for danger or trouble

Morse code (MORS kode) a communication system that uses light or sound in patterns of dots and dashes to represent letters and numbers

navigate (NAV-i-gate) to sail along or across

ocean liner (OH-shuhn LYE-nur) a ship that runs on a regular schedule from one seaport to another

saloon (suh-LOON) a large social lounge on a passenger ship

stateroom (STATE-room) a private cabin or compartment with sleeping accommodations on a ship or train

Index

Page numbers in **bold** indicate illustrations.

About the Author

Laura McClure Anastasia has been fascinated with the *Titanic* ever since her parents took her to an exhibit about the ill-fated ship when she was a teen. She pored through primary and secondary accounts of the tragedy to research this book. A graduate of the University of Pennsylvania, Laura has worked in educational publishing for two decades. She is an editor for a national news magazine for kids. She loves that her work gives her an excuse to ask questions and learn more about the world around her every single day. Laura lives in Bradenton, Florida, with her two fun children and her very adorable dog.

Photos ©: cover main: Gary Hanna; cover top: The Titanic Collection/UIG/Bridgeman Images; cover bottom: Mary Evans Picture Library Ltd/age fotostock; back cover: Lordprice Collection/Alamy Images; 3: Jonathan Barry. All Rights Reserved 2022/Bridgeman Images; 4: Bettmann/Getty Images; 5 top: Pictures from History/Bridgeman Images; 5 bottom: Everett Collection/age fotostock; 6–7: Dorling Kindersley/Science Source; 8–9: Harley Crossley. All Rights Reserved 2022/Bridgeman Images; 8 bottom: Jim McMahon/Mapman ®; 10: Shutterstock; 11: Look and Learn/Bridgeman Images; 12: Agence Roger Viollet/The Granger Collection; 13: ARCHIVIO GBB/Alamy Images; 14: Paul Quezada-Neiman/Alamy Images; 15 background: Historia/Shutterstock; 15 top: Prismatic Pictures/Bridgeman Images; 15 bottom: Haldridge/Bournemouth News/Shutterstock; 16–17: Prismatic Pictures/Bridgeman Images; 18 left: United Archives/Carl Simon/Bridgeman Images; 18 right: The Granger Collection; 19: Patrice Cartier/Bridgeman Images; 20 top: National Maritime Museum, Greenwich/Bridgeman Images; 20 bottom left: Science & Society Picture Library/Getty Images; 20 bottom right: The Granger Collection; 21: Prismatic Pictures/Bridgeman Images; 22: Pictures from History/Bridgeman Images; 23: Bettmann/Getty Images; 24 bottom: Roger-Viollet/TopFoto; 24 top: The Titanic Collection/UIG/Bridgeman Images; 25: Illustrated London/Mary Evans Picture Library Ltd/age fotostock; 26–27 background: Shutterstock; 26 foreground: The Print Collector/Getty Images; 28: Archiv Carl Simon/UA/Bridgeman Images; 29 left: Jonathan Barry. All Rights Reserved 2022/Bridgeman Images; 29 right: Everett Collection Inc./age fotostock; 30: Bettmann/Getty Images; 31: Gary Hanna; 32–33: Hulton Archive/Getty Images; 33 top: United Archives/Carl Simon/Bridgeman Images; 33 bottom: United Archives/Carl Simon/Bridgeman Images; 34–35: Kollected; 36 left: Roger-Viollet/TopFoto; 36 center: Archiv Carl Simon/UA/Bridgeman Images; 36 right: Hulton Archive/Getty Images; 37 left: Bettmann/Getty Images; 37 center: Gary Hanna; 37 right: Everett Collection/age fotostock; 38, 39: United Archives/Carl Simon/Bridgeman Images; 40–41 background: Lordprice Collection/Alamy Images; 40 bottom: The Titanic Collection/UIG/Bridgeman Images; 41 bottom: Photo Researchers/Science History Images/Alamy Images; 42–43: Kollected; 44: Roger-Viollet/TopFoto.